Process Intensification 2020

Topical Conference at the 2020 AIChE Spring Meeting and 16th Global Congress on Process Safety

Online
17-21 August 2020

ISBN: 978-1-7138-2336-0

Printed from e-media with permission by:

Curran Associates, Inc.
57 Morehouse Lane
Red Hook, NY 12571

AIChE
120 Wall Street, FL 23
New York, NY 10005-4020

Phone: (800) 242-4363
Fax: (203) 775-5177

www.aiche.org

Additional copies of this publication are available from:

Curran Associates, Inc.
57 Morehouse Lane
Red Hook, NY 12571 USA
Phone: 845-758-0400
Fax: 845-758-2633
Email: curran@proceedings.com
Web: www.proceedings.com

TABLE OF CONTENTS

591338 Supercritical Fluid Extraction for Process Intensification of Biological Processes

Abstract:

The energy efficiency of many biological processes is limited by recovery of dilute fermentation products. Ethanol is an exception to this rule, as it can be tolerated by microbes at concentrations greater than 10 wt%. Other products are much more toxic, which limits their endpoint concentrations and makes energy intensive their recovery. While distillation remains the separation technology of choice when feasible, energy efficient alternatives are required for instances where endpoint concentrations are insufficient for distillation, when products are insufficiently volatile, or when products are thermally sensitive. One promising technology, which has been known since the 1980s but incompletely developed, is supercritical fluid extraction. Supercritical fluid extraction combines elements of distillation with liquid-liquid extraction. The extraction solvent, typically carbon dioxide or a light hydrocarbon, is present in its supercritical state, meaning that the thermodynamic driving force is based on solute partitioning – similar to liquid-liquid extraction. Subsequent to extraction, the target molecule can be recovered in high purity simply by decreasing the pressure, meaning that supercritical fluid extraction does not require downstream separation steps – a feature that is held in common with distillation. Ideal targets should be amenable to biomanufacturing, but with titers limited by end product inhibition. They should be water soluble at concentrations less than their inhibition limit and partition favorably into one or more supercritical solvents and out of aqueous solution. Pressure and temperature can be adjusted to fine tune the separation. Butanol is used as a test case to show dramatic, >10-fold decreases in energy requirements for separation. New advances that combine microbes that are tolerant of supercritical fluids and capable of bioproduction of desirable target molecules hold promise for even greater improvements in energy efficiency, as demonstrated by a recent study using the newly identified strain *Bacillus megaterium SR7*. Integrated fermentation and extraction benefits from optimized process design including solvent recycling and co-production of carbon dioxide at high pressure. Future studies can investigate mass transfer in the supercritical fluid-water environment. Future development work can focus on reducing costs.

591331 Optimizing Cyclic PSA/TSA Conditions for Ammonia Separation with Supported Metal Halides

Abstract:

The recent push towards reduction of carbon dioxide emissions has been the main reason for finding alternatives to the conventional Haber-Bosch ammonia production process. Although well-optimized through a century of research and development, the high operating pressure of the Haber-Bosch process is economically demanding. Alternative pathways to produce green ammonia introduce renewable energy (mostly wind-generated) as a base for ammonia production with the downside being double the cost. This shortcoming is mitigated by operating the process at lower pressure by replacing the condenser with an absorber containing metal halides capable of better ammonia separation – using the reaction-absorption process.[1–4] The application of absorption for ammonia separation is well studied; yet, there is no information about the rates of ammonia release from the metal halide absorbents.[5,6]

In this presentation, we offer one of the initial efforts to study ammonia release from different metal halide absorbents. The target is to better understand the ammonia release rates and absorbent working capacity under the cyclic pressure and temperature swing absorption conditions (PSA/TSA). A lab-scale absorber packed with supported and unsupported materials is used to observe ammonia breakthrough and release times for each cycle. Ammonia release at specific temperatures and pressures is recorded in carefully controlled experiments. Absorption tests were conducted at 40 psi with temperatures in the range of 25-400 °C. Results show a viable path to achieve reproducible breakthrough curves while achieving ammonia concentrations above 90%. The effect of the sweep gas during the desorption is also studied with the purpose of finding the optimal values for sweep time and volumetric flow. These findings will guide us to further optimize the small reaction-absorption ammonia production process and explore the strategies for further scale-up.

References:

(1) Malmali, M.; Wei, Y.; McCormick, A.; Cussler, E. L. Ammonia Synthesis at Reduced Pressure via Reactive Separation. *Ind. Eng. Chem. Res.* **2016**, *55* (33), 8922–8932.

(2) Malmali, M.; Reese, M.; McCormick, A. V.; Cussler, E. L. Converting Wind Energy to Ammonia at Lower Pressure. *ACS Sustain. Chem. Eng.* **2018**, *6* (1), 827–834.

(3) Reese, M.; Marquart, C.; Malmali, M.; Wagner, K.; Buchanan, E.; McCormick, A.; Cussler, E. L. Performance of a Small-Scale Haber Process. *Ind. Eng. Chem. Res.* **2016**, *55* (13), 3742–3750.

(4) Cussler, E.; McCormick, A.; Reese, M.; Malmali, M. Ammonia Synthesis at Low Pressure. *J. Vis. Exp.* **2017**, No. 126, e55691–e55691.

(5) Wagner, K.; Malmali, M.; Smith, C.; McCormick, A.; Cussler, E. L.; Zhu, M.; Seaton, N. C. A. Column Absorption for Reproducible Cyclic Separation in Small Scale Ammonia Synthesis. *AIChE J.* **2017**, *63* (7), 3058–3068.

(6) Malmali, M.; Le, G.; Hendrickson, J.; Prince, J.; McCormick, A. V.; Cussler, E. L. Better Absorbents for Ammonia Separation. *ACS Sustain. Chem. Eng.* **2018**, *6* (5).

591297 Material Selection in Process Intensification: Application to Extractive Separation Systems

Abstract:

Extractive separation, a classic process intensification system, offers the solution to overcome the physical equilibrium at azeotropic point by introducing an extra mass separating agent (i.e., entrainer) [1-2]. The material selected plays a key role due to: (i) its solubility properties to enhance the system's mass transfer driving force to drive separation, and (ii) its hazardous properties (e.g., toxicity, flammability) for environmental and safety considerations. Despite recent advances in the synthesis of process intensification systems leveraging phenomenological representation methods to deliver potentially "out-of-the-box" process solutions [3-5], the role of (advanced) materials in process synthesis/intensification has not been sufficiently addressed in open literature [6].
In this work, we propose a systematic approach for material selection, process synthesis, integration, and intensification based on recent extensions of the phenomena-based Generalized Modular Representation Framework (GMF) [7-8]. Herein, processes are represented as aggregated multifunctional mass/heat exchange modules with which intensification possibilities can be discovered without a pre-postulation of equipment or flowsheet configurations. Driving force constraints, derived from total Gibbs free energy change, are employed to characterize mass/heat transfer feasibility by exploiting the general thermodynamic space, and thus result in a more compact modular representation of the chemical systems. Material selection is achieved by incorporating rigorous thermodynamic models (e.g., NRTL) in GMF representation to capture the nonideal liquid mixture properties as well as to assess material performance to facilitate separation. Mass and/or heat integration are also systematically integrated in the superstructure formulation of the modular network to reduce energy consumption, resource utilization, and waste production. The resulting synthesis problem is formulated as a mixed integer nonlinear programming optimization problem (MINLP), where material selection, process synthesis, integration, and intensification can be simultaneously accounted for with respect to process profitability. An extensive case study concerning enthanol/water extractive separation is presented to showcase the proposed approach, with two entrainer candidates examined: a conventional entrainer ethylene glycol (EG) and an ionic liquid entrainer 1-ethyl-3-methyl-imidazolium acetate ([EMIM][OAc]). Extensions of the framework to include process safety and environmental considerations will also be discussed [8]. 1. Tian, Y., Demirel, S. E., Hasan, M. M. F., & Pistikopoulos, E. N (2018). An Overview of Process Systems Engineering Approaches for Process Intensification: State of the Art. Chemical Engineering and Processing: Process Intensification, 133, 160-210. 2. Zhou, T., Song, Z., Zhang, X., Gani, R., & Sundmacher, K. (2019). Optimal Solvent Design for Extractive Distillation Processes: A Multi-objective Optimization based Hierarchical Framework. Industrial & Engineering Chemistry Research. 3. Demirel, S. E., Li, J., & Hasan, M. F. (2017). Systematic process intensification using building blocks. Computers & Chemical Engineering, 105, 2-38. 4. da Cruz, F. E., & Manousiouthakis, V. I. (2016). Process intensification of reactive separator networks through the IDEAS conceptual framework. Computers & Chemical Engineering, 105, 39-55. 5. Tula, A. K., Babi, D. K., Bottlaender, J., Eden, M. R., & Gani, R. (2017). A computer-aided software-tool for sustainable process synthesis-intensification. Computers & Chemical Engineering, 105, 74-95. 6. Stankiewicz, A., & Yan, P. (2019). 110th Anniversary: The Missing Link Unearthed: Materials and Process Intensification. Industrial & Engineering Chemistry Research. 7. Papalexandri, K. P., & Pistikopoulos, E. N. (1996). Generalized modular representation framework for process synthesis. AIChE Journal, 42(4), 1010-1032. 8. Tian, Y., & Pistikopoulos, E. N. (2018). Synthesis of Operable Process Intensification Systems — Steady-State Design with Safety and Operability Considerations. Industrial & Engineering Chemistry Research, 58(15), 6049-6068.

590190 Data-Driven Approach to Systematic Process Intensification Using Building Block Superstructure

Abstract:

Process intensification (PI) involves improvements in process performance that can lead to substantially smaller, cleaner, safer, and more energy-efficient technologies [1, 2]. However, identification of intensified solutions at the conceptual design stage is a challenge as there can be a plethora of possible process configurations. Also known as superstructure optimization, optimization-based process synthesis can be used to identify the optimal process configuration. However, traditional optimization-based approach requires all process configurations and alternatives to be specified beforehand with fixed connectivity and equipment types. Such pre-postulated superstructure may not exist in some cases; thus, the optimal intensified process cannot be identified [3].

A building block superstructure has been proposed recently to overcome the aforementioned challenge [4]. Unlike the conventional superstructure approach, the building block superstructure does not require *a priori* postulation of process alternatives. Instead, each building block is used to represent a fundamental constituent of a unit operation, which can then be combined together to represent physicochemical phenomena. These building blocks can be positioned on a two-dimensional grid, and the optimal intensified flowsheet can be automatically generated by solving a mixed-integer nonlinear programming (MINLP) model [5]. However, as these building blocks are represented using a set of algebraic equations describing complex thermodynamic and material and energy transfer, it contains several non-convex equations, which increases the complexity of the final optimization model. Therefore, locating an optimal solution still remains a challenge [6].

In this work, we propose the use of data-driven approach for the design of a building block superstructure. Specifically, instead of relying on a complicated set of algebraic equations describing physical phenomena, we construct accurate and tractable machine learning-based process models to reduce the complexity of the building block superstructure model [7]. In particular, we use an iterative machine learning-based modeling and optimization algorithm, which involves: 1) constructing an efficient design of experiments, 2) constructing machine learning-based process models, and 3) formulating and solving the machine learning-based MINLP problem. A case study on a reactive separation system with non-ideal thermodynamics will be presented, and known process information and constraints will be considered during the design stage to identify a realistic intensified design.

1. Stankiewicz, A. and J.A. Moulijn, *Process Intensification: Transforming Chemical Engineering*. Vol. 96. 2000. 22-33.
2. Tian, Y., et al., *An overview of process systems engineering approaches for process intensification: State of the art.* Chemical Engineering and Processing - Process Intensification, 2018. **133**: p. 160-210.
3. Demirel, S.E., J. Li, and M.M.F. Hasan, *A General Framework for Process Synthesis, Integration, and Intensification.* Industrial and Engineering Chemistry Research, 2019. **58**(15): p. 5950-5967.
4. Li, J., S.E. Demirel, and M.M.F. Hasan, *Process synthesis using block superstructure with automated flowsheet generation and optimization.* AIChE Journal, 2018. **64**(8): p. 3082-3100.
5. Demirel, S.E., J. Li, and M.M.F. Hasan, *Systematic process intensification using building blocks.* Computers & Chemical Engineering, 2017. **105**: p. 2-38.
6. Boukouvala, F., M.M.F. Hasan, and C.A. Floudas, *Global optimization of general constrained*

grey-box models: new method and its application to constrained PDEs for pressure swing adsorption. Journal of Global Optimization, 2017. **67**(1): p. 3-42.
7. Kim, S.H. and F. Boukouvala, *Machine learning-based surrogate modeling for data-driven optimization: a comparison of subset selection for regression techniques.* Optimization Letters, 2019.

591511 Electrification, Process Intensification, and Artificial Intelligence for Modular and Improved Chemical Manufacturing

Abstract:

There is renewed interest in the chemistry and processing of small molecules, such as carbon dioxide, CO_2, ammonia, NH_3, and small hydrocarbons (methane, ethane and propane). The large energy demand and associated large production of CO_2 in ammonia production, the conversion of CO_2 to chemicals and fuels, and the revolution of shale gas in the U.S., and thus of small hydrocarbons, are changing the global energy and chemicals terrain and are creating new opportunities for research and manufacturing. Common to the small molecules are the challenges of activating chemical bonds, the large energy demand, the equilibrium limitations, the loss of selectivity, and catalyst deactivation. In addition, processing some of these resources may require distributed manufacturing, whose economic viability is uncertain. These challenges require understanding of the fundamental chemistry and development of new catalysts and processes. This talk will overview these emerging trends and present examples from chemical activation using electrification, such as microwaves, coupled with other process intensification concepts to allow scalability. The role of data science in process design will also be discussed.

590197 Advanced Integrated Techniques for Quantifying Kinetics and Transport Parameters in Multiphase Reactors

Abstract:

Multiphase reactors and multiphase flow systems have found extensive related to energy, environmental, chemicals, petroleum and variety of products processes. Multiphase flow reactors are complex due to the intricate interactions and reactions among difference phases. Therefore, sophisticated techniques that integrate measurements in a novel way of various hydrodynamics and transport parameters and their integration with kinetics are required to quantify and visualize these systems to provide benchmarking data for validating CFD and models. These advances should provide the mechanistic frame work to properly scale up the results from lab to commercial scales.

In our laboratory, various types of multiphase systems such as slurry/basket reactors and packed bed reactors under high temperature and high pressure for various industrial and environmental processes have been studied by developing advanced measurement techniques and validating CFD simulations which enable process intensification and facilitating the implementation of new mechanistic scale-up methodologies. In general, these techniques can be classified into two categories: radioisotope techniques and non-radioisotopes techniques. The radioisotope techniques include radioactive particle tracking (RPT), dual source gamma ray tomography (DSCT), gamma ray densitometry (GRD) for 3D flow field, velocity and turbulent parameters, phases distribution and flow pattern identification measurements. The non-radioisotope techniques include 4-point optical probe for bubble dynamics, optical probes for gas-solid dynamics, optical probe for local mass transfer, optical probe for local liquid and gas velocities, holdup and their time series distribution in packed beds, gas tracer technique for global mass transfer, heat transfer probe, pressure transducers, and others.

591298 Process Intensified Conversion of Stranded Energy Resources to Value-Added Liquid Products for Distributed Production

Abstract:

Over 200 billion ft^3/year of natural gas at remote production sites is flared in the U.S. due to the limitations in pipeline transportation capacity. This wastes valuable resources, increases air emissions, and burns profits. Similarly, renewable electricity (wind and solar) in some disadvantaged locations becomes stranded due to the limitations in power transmission capacity and intermittent nature of the renewable electivity. This presentation introduces the applications of microwave catalysis for the conversion of these stranded energy resources into liquid transportable chemicals such as aromatics and ammonia. Microwave irradiation has shown a profound impact on catalyzed and uncatalyzed gas-solid reactions. In particular, it has been demonstrated that microwave-specific effects can manifest themselves through the enhancement of reaction rates, changes in the position of equilibria and the distribution of products.

Methane must be activated to a high-energy transition state before it can further react to produce stable products. The high reaction temperatures (>700°C) required to activate methane's strong C-H bonds usually results in rapid and extensive coke formation that blocks catalyst micropores, deactivating the catalyst. The gaps in scientific understanding of the direct, non-oxidative methane conversion process preclude the design of highly selective catalysts and reactors to make desired higher value products. Technical hurdles include catalyst stability and thermodynamic limitations that constrain the desired product yields. In this study, methane conversion reaction is enhanced with microwave irradiation to demonstrate microwave reaction chemistry at catalyst-reactant interface for selective C-H bond activation

Stranded electricity can be converted to liquid ammonia under ambient pressure and 320°C using water and air as feedstock. Conventional ammonia synthesis process, i.e. Haber-Bosch process, cannot be scaled down economically due to the high temperature and high pressure operation and the intermittent nature of renewable electricity that requires frequent startup and shutdown. Catalytic ammonia synthesis was conducted under two microwave irradiation scenarios: microwave and microwave plasma. Other than selective activation of dinitrogen to metastable radicals, the most obvious advantage that microwave irradiation affords in driving a heterogeneously catalyzed reaction is the ability to locally heat the catalytic sites. Many industrial processes utilizing heterogeneous catalysts are high-temperature processes wherein both components of the reaction (i.e., catalyst and medium) are heated to the temperature required for the reaction to occur. Specifically discussed in this presentation are the effects of electromagnetic properties of catalysts, microwave frequency and microwave energy absorbed on the conversion of nitrogen and ammonia yield.

Technoeconomic analysis was carried out to compare microwave catalytic processes with the state-of-the art. The comparisons are made to focus on process intensification elements such as energy efficiency, energy productivity and capital cost reduction. Sensitivity analysis was made to evaluate various scenario and different plant sizes. The resulting engineering innovation includes plant siting flexibility due to smaller scale, lower capital and operating costs, higher energy efficiency, lower carbon footprint, no or low environmental emissions.

The potential for a dramatic transformation in the chemical process industries is emerging from the rapid development of process intensification. This can potentially yield substantial economic and environmental benefits in combination with advanced manufacturing by significantly reducing the size of chemical process plants. This presentation aims to revolutionize the chemical process industries by enabling the development of intensified and, where appropriate, modular process plants.

591336 Process Intensification of Ternary Distillation Using Dynamic Optimization Method and Data-Driven Approach

Abstract:

Distillation is the dominant separation technology for liquid mixtures in chemical industry. Studies revealed that over 40,000 distillation columns in operation in the US are responsible for more than 90% of purification needs [1]. The thermodynamic efficiency of distillation can be as low as 10% [2], and the operation of distillation can account for half of the energy consumed by chemical plants [3].

Process intensification (PI) is by now an accepted means for improving the energy efficiency of distillation column (sequences). Through the implementation of innovative design changes, process intensification significantly lowers both capital and operating costs. However, tight design integration/integration comes with control challenges: a reduced number of control degrees of freedom and a more constrained operation window compared to conventional columns [4].

In this paper, we rely on a different approach to process intensification, implemented in the temporal domain, which we refer to as "dynamic process intensification (DPI)." DPI aims to improve process efficiency by exploiting and favorably manipulating in the time domain the nonlinear dynamics of existing columns [5-7]. The principle of DPI consists of imposing operational instead of design changes, aiming to meet product quality constraints while minimizing energy per production rate through implementing a transient (periodic) operating pattern. Thus far, such patterns were defined in terms of switching between a finite (generally, two) fixed operating points. In this paper, we introduce a novel approach based on dynamic optimization. We introduce a data-driven approach for learning the DPI optimization-relevant dynamics of a distillation process from its operational history and present a dynamic optimization-based DPI framework. We show that, among others, Hammerstein-Wiener (HW) models that have fewer parameters to regress and can be trained with a considerably smaller amount of data can accurately capture the dynamics of the distillation plant along with its control system [8-9].

We will apply this conceptual framework to an extensive case study separating cyclohexane, toluene and m-xylene mixture. Meeting all quality constraints, DPI can offer considerable energy savings compared to steady-state operation of the same column.

References

[1] Humphrey, J. L. Separation processes: playing a critical role. *Chemical Engineering Progress,* **1995**, 91.na

[2] Humphrey, J. L.; Seibert, A.; Koort, R. Separation Technologies: Advances and Priorities,

tech. rep., Office of Industrial Technologies, Washington DC, **1991**.

[3] Isopescu, R.; Woinaroschy, A.; Draghiciu, L. Energy reduction in a divided wall distillation column. *Revista de Chimie* **2008**, *59*, 812-815.

[4] Schembecker, G.; Tlatlik, S. Process Synthesis for Reactive Separations, *Chem. Eng. Process*, **2003**, *42*, 179-189.

[5] Baldea, M.; Edgar, T. F. Dynamic Process Intensification, Dynamic Process Intensification, *Current Opinion in Chemical Engineering*, **2018**, *22*, 48-53.

[6] Yan, L.; Edgar, T. F.; Baldea, M. Dynamic Process Intensification of Binary Distillation Based on Output Multiplicity, *AIChE J*, **2019**, *65(4),* 1162-1172.

[7] Yan, L.; Edgar, T.; Baldea, M. Dynamic Process Intensification of Binary Distillation via Periodic Operation, *Industrial & Engineering Chemistry Research,* **2019**, *58(15)*, 5830-5837.

[8] Richard C Pattison, Cara R Touretzky, Ted Johansson, Iiro Harjunkoski, and Michael Baldea. Optimal process operations in fast-changing electricity markets: framework for scheduling with low-order dynamic models and an air separation application. *Industrial & Engineering Chemistry Research*, 55(16):4562–4584, 2016

[9] Ian Goodfellow, Yoshua Bengio, and Aaron Courville. Deep learning. *MIT Press*, Cambridge, MA, 2016.

589766 Process Intensification of CO_2 Absorption Using an Additively Manufactured Intensified Packing Device

Abstract:

A multifunctional packing device has been developed, characterized, and tested for process intensification of CO_2 absorption by aqueous amines in packed columns. The multifunctional device is an additively manufactured packing element that integrates effective contact of phases and heat exchange in multiphase systems, and can be employed to vent excess heat from an exothermic reactive system or supply thermal energy to an endothermic chemical reaction. The impact of excess heat removal by the device on the absorption of CO_2 by aqueous solutions of monoethanolamine is investigated. The design process of the device and the validation of core hydrodynamic metrics will be discussed. Heat and mass transfer testing demonstrated that the device was effective at removing excess heat from the reaction and, furthermore, that cooling could successfully enhance capture of CO_2 by the amine solvent. The magnitude of the enhancement was found to depend strongly on operating conditions, in particular flow rates of CO_2, air, and solvent concentration. This device represents a successful example of the role of additive manufacturing in the intensification of chemical processes.

590385 An Efficient and Reliable Numerical Method for Computing Coupled Multicomponent Fluxes for Complex Mixtures across Asymmetric Membranes

Abstract:

An Efficient and Reliable Numerical Method for Computing Coupled Multicomponent Fluxes for Complex Mixtures Across Asymmetric Membranes

Dylan Weber[1], Ronita Mathias[1], Ryan Lively[1], Chau-Chyun Chen[2], Joseph Scott[1†]

1 Department of Chemical and Biomolecular Engineering, Georgia Institute of Technology, Atlanta, GA 30332, USA

2 Department of Chemical Engineering, Texas Tech University, Lubbock, TX, 79409, USA.

Corresponding author; Email: joseph.scott@chbe.gatech.edu

Abstract

Membranes offer a promising means to intensify many industrial separation processes that are currently achieved using distillation and other thermal separation technologies. Historically, key limitations in the design and synthesis of high-performance membrane materials has posed the most significant barrier to the adoption of membrane processes in industry. However, with steady improvements in membrane material properties and module construction, the lack of reliable membrane modeling and simulation tools is emerging as a major barrier in its own right. In particular, widespread industrial adoption of membrane processes as an alternative to conventional thermal separations is unlikely to occur in the absence of (i) reliable predictive models for membrane performance when confronted with complex multicomponent streams, (ii) methods and databases for parameterizing such models using pure component data and limited multicomponent measurements, and (iii) efficient and reliable numerical solution methods that are compatible with standard process flowsheeting software and enable technoeconomic comparison of membrane and non-membrane processes on a consistent basis.

In this presentation, we will describe a novel numerical solution procedure for computing transmembrane fluxes for non-ideal multicomponent mixtures in contact with an asymmetric membrane. Numerous relevant applications of membrane separations such as water purification, carbon capture, hydrogen separation, olefin/paraffin separation, and benzene derivative concentration include process streams that are multicomponent nonideal mixtures. Along with that there are many different types of membrane material being used. From highly selective zeolites, carbon nanotubes, ionic polymers, mixed-matrix MOF and polymeric, graphene-oxide, and specialty polymers. Many of the current approaches for industrial practice adopt the permeability and linear driving force model to describe permeation across the selective layer. These models are applicable in many cases where the sorption is ideal and uncoupled transport can be assumed. However, this explanation of the system when trying to model many nonideal components and interacting permeants is not correct. That is why our group is looking to employ a feasible rigorous solution of the Maxwell-Stefan equations to describe the multicomponent mass transport. Past contributions from experts in these equations such as Rajamani Krishna, Benny Freeman, or Pavel Izák employ approximate solutions by evaluating terms at average compositions, assume the permeate composition is zero to allow for an approximate solution, or use finite differences to approximate the differential terms. These studies are also restricted to a ternary system with two permeates and a membrane phase. No work has been done to have an effective and reliable numerical method for a system with any number of components.

As a brief overview, the Maxwell-Stefan (MS) approach towards mass-transport is based on irreversible thermodynamics by balancing the driving forces any component from a mixture experienced in the form of the chemical potential with the collective friction forces experienced by transporting with other components in the mixture. The Maxwell-Stefan diffusivities can be physically interpreted as the inverse friction or drag coefficients of a binary interaction between two permeating molecules or the permeant and the membrane phase. The main advantages of this transport model compared to the Fickian model is the incorporation of driving forces other than a concentration gradient, but namely MS diffusivities are not dependent on the system and are an intrinsic property of the binary component or membrane pair. Note, these diffusivities can still be functions of temperature, pressure, and composition. The difference is the general applicability from single or binary permeation data to the multicomponent system with many nonidealities accounted for. Then with the nature of many nonidealities, there are many problems in solving such systems of equations. The problems associated with the complete solution of the set of MS equations are the numerical methods associated with solving the matrix formulation of the set of coupled ordinary differential equations. Our research goal is to provide a working rigorous solution method that converges to physically meaningful solutions with only feed stream inputs, a thermodynamic activity model with required parameters, and single component membrane diffusivities (in some cases binary component diffusivities). This will provide a working toolset for the practicing chemical engineer to design these systems without having to be an expert in the underlying transport equations and allow for rapid plant design prototyping and screening of various process units. In this work we consider the Maxwell-Stefan as the general transport model, and the multicomponent Flory-Huggins equation to describe the activity of the permeants in the polymer phase. The interesting note is this contribution is applicable to any membrane system that can be described by the Maxwell-Stefan equations.

590118 Ethylene Production Using Oxidative Dehydrogenation: Effects of Alternative Separation Technologies on Process Safety & Economics

Abstract:

Summary

Previous work from this group utilized a HYSYS simulator to predict the performance of an ethylene oxidative dehydrogenation (ODH) production facility using alternative diluents and separation technology not currently used in the standard industrial process [*1-3*]. An ODH process system has been modeled with ethane being fed along with air into a reactor with simultaneous mixing with pure methane to lower the product gas temperature. The methane feed was later separated from the other product gases and eventually fed into a recycle stream. This process was fairly expensive due to the use of a demethanizer within the process recycle stream. Safety and operational improvements for the ODH process was another aspect of this analysis that yielded interesting results. By the inclusion of storage vessels before and after the ODH reactor, the reactor unit can operate even in the event of an upstream or downstream failure, resulting in less downtime for the process.

The current work investigated the development of alternative separation technologies for the production of ethylene for the oxidative dehydrogenation (ODH) reaction of ethane using a HYSYS simulation with economic and safety analyses. The application of polymer-membrane separation and non-cryogenic distillation alternatives, such as bottom-flash, heat-pump assisted and low-temperature distillation, were simulated and evaluated individually and in-tandem with membrane separation. The effects of alternative diluents, process flammability issues, and economic analysis results will also be discussed.

Literature Cited:

1.Maffia GJ, Gaffney AM, Mason OM, Techno-economic analysis of oxidative dehydrogenation options, Top. Catal. (2016), http://dx.doi.org/10.1007/ s11244-016-0677-9 (2016, August 16).

2.Baroi C, Gaffney AM, Fushimi R, Process economics and safety considerations for the oxidative dehydrogenation of ethane using M1 catalyst, Catal. Today (2017), http://dx.doi.org/10.1016/j.cattod.2017.05.041 R (2017, May 11).

3.Gao, Y., Neal, L., Ding, D., Wu, W., Baroi, C., Gaffney, A.M., Li, F., Recent Advances in Intensified Ethylene Production: A Review, ACS Catalysis (2019), http://10.1021/acscatal.9b02922 (2019, August 12).

614862 Energy Efficient Separations of Olefins and Paraffins through a membrane II

Abstract:

Update on the Separation of Olefin-Paraffin Mixtures with Optiperm™ Customized Amorphous Fluoropolymer Membranes

Abstract: The separation of light olefins from paraffins is one of the most energy intensive in the world. The state of the art separation method is distillation, a technology that has remained unchanged over decades. Membranes hold the promise of a more modular, energy efficient and capital efficient method for separation. However, they have historically been hampered by low performance and stability.

Compact Membrane Systems has developed a novel membrane material using a silver embedded fluoropolymer material to transport olefins across the membrane. This material has demonstrated over 2 years of longevity in laboratory operation and has shown high selectivity and permeance under realistic operating conditions. Additionally Optiperm membranes were trialed at the Delaware City Refining Company in over 300 days of field operations validating the robustness and capability of the technology in an operating plant environment.

During the past year, CMS has developed and installed a laboratory recycle test systems to allow for the continuous testing of spiral wound modules with areas of ~20 square feet. The recycle test unit enables CMS to test larger scale membrane modules under field operating conditions using actual process streams from end users. Furthermore the recycle rig design incorporates our innovative design for hydrating the process feed within the membrane housing, significantly reducing the complexity of the membrane systems operations.

We will share the latest leanings from our recycle rig testing as well as provide updates on the scale up of our flat sheet spiral wound membrane cartridges. We will also discuss the current state of our hollow fiber membrane configuration development and the progress on our joint project with Dow Chemical through RAPID, a manufacturing institute funded by the Department of Energy. The goal of the work with Dow is to produce a chemical grade propylene from a process feed stream with a propylene concentration of greater than 60%.

CMS will also discuss results on laboratory testing of C_4 streams. This includes mixed gas separation of n-butane and 1-butene with permeances between 200-500 GPU and selectivity greater than 20. We will share results for a number of operating conditions and will highlight several exciting C_4 separation applications including those with isomer streams. In addition, longevity results will be shared demonstrating the stability of the membrane in C_4 separations.

589680 Radio Frequency Driven Catalytic Reactors for Portable Green Chemistry

Abstract:

Heterogeneous catalytic reactors account for roughly 80% of all chemical manufacturing processes,[1] and are often driven by external heating of the catalyst via fuel-fired furnaces or utility steam. Traditional methods relying on combustion of fuel often lead to greenhouse emissions, limit distributed chemical production due to economy of scale, require additional infrastructure and large reactor size to accommodate combustion zones and insulation around the reactor. They also have thermal gradients in reactors which often compromises performance, utilization, lifetime, and selectivity of the catalyst.

The volumetric electric heating methods and modular reactors for power-to-chemicals route can pave a path towards distributed manufacturing and reduced greenhouse gas emissions when electricity is harnessed from renewable energy sources. Uniform volumetric heating of catalyst can also improve catalyst utilization, avoid homogeneous side reactions, and improve reactor portability.[2] Since the first published report of microwave organic synthesis, [3-4] microwave heating has been studied for various endothermic catalytic reactions.[5] However, high power microwaves (> 2.45 GHz frequency) are limited due to temperature hotspots, runaway reactions, penetration depth, reflection losses, and stringent safety exposure limits.

This study employs a multidisciplinary approach to make portable reactors by using novel materials like carbon nanotubes (CNTs) and silicon carbide (SiC) fibers as additives in catalyst. We utilize interaction of these materials with radio frequency fields (1 MHz-300 MHz) to selectively heat catalytic reactors and drive a reaction. A proof-of-concept is demonstrated for methanol steam reforming reaction using platinum as a catalyst. The RF heating response of CNT/platinum /alumina and SiC fiber/platinum were investigated for varying power using different kinds of noncontact RF applicators. The conversion of methanol for different reaction temperatures was compared to conventional ovens. This power to chemical method has application in modular reactors for on-site and on-demand production of chemicals using electric power.

(1) Fechete, I.; Wang, Y.; Védrine, J. C. The past, present and future of heterogeneous catalysis. *Catalysis Today* **2012,** *189* (1), 2-27, DOI: https://doi.org/10.1016/j.cattod.2012.04.003.

(2) Wismann, S. T.; Engbæk, J. S.; Vendelbo, S. B.; Bendixen, F. B.; Eriksen, W. L.; Aasberg-Petersen, K.; Frandsen, C.; Chorkendorff, I.; Mortensen, P. M. Electrified methane reforming: A compact approach to greener industrial hydrogen production. *Science* **2019,** *364* (6442), 756-759, DOI: 10.1126/science.aaw8775.

(3) Gedye, R.; Smith, F.; Westaway, K.; Ali, H.; Baldisera, L.; Laberge, L.; Rousell, J. The use of microwave ovens for rapid organic synthesis. *Tetrahedron letters* **1986,** *27* (3), 279-282.

(4) Tse, M. Y.; Depew, M. C.; Wan, J. K. S. Applications of high power micro wave catalysis in chemistry. *Research on Chemical Intermediates* **1990,** *13* (3), 221-236, DOI: 10.1163/156856790x00102.

(5) Ramirez, A.; Hueso, J. L.; Abian, M.; Alzueta, M. U.; Mallada, R.; Santamaria, J. Escaping undesired gas-phase chemistry: Microwave-driven selectivity enhancement in heterogeneous catalytic reactors. *Science Advances* **2019,** *5* (3), eaau9000, DOI: 10.1126/sciadv.aau9000.

Membrane Water Gas Shift Reactor Design: Applied to 10 MTPD Modular Ammonia Plant.

By Richard B. Strait

Abstract:

Extensive research is being performed to develop membrane reactors for hydrogen production. Much of the research is focused on the Steam Methane Reforming (SMR) reaction and to a lesser extent the Water Gas Shift (WGS) reaction. In parallel we have been designing a 10 metric ton per day (MTPD) modular ammonia plant which can be transported to sight as cargo and set up to operate using local feedstock to produce fertilizer for local consumption. If a low cost reliable membrane SMR or WGS or even Ammonia Converter were available they would allow for smaller equipment that did more making the ammonia process more cost effective and suitable for compact, small modular plants.

This paper focuses on one of these membrane reactors, the WGS. The basic design will be presented and the positive aspects of the design along with the challenges as they apply to a modular Ammonia Manufacturing Box (AMB) producing 10 MTPD next to the farmer's field.

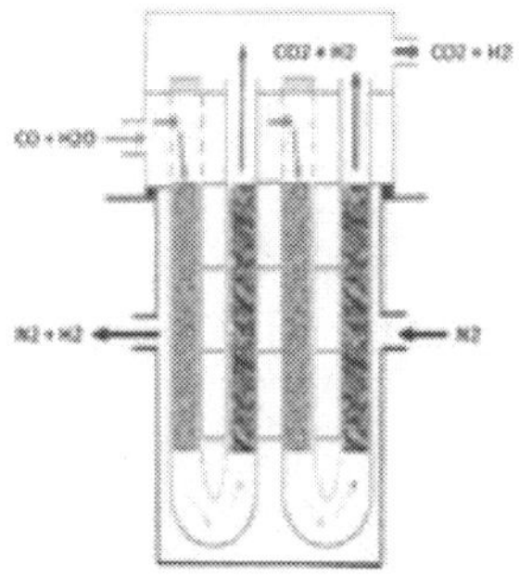

The mechanical design is discussed including:

- Membrane materials selection, cost and reliability
- Construction, maintenance, turnaround and emergency operation.

The process design is discussed including:

- WGS Reactor and catalyst performance
- Hydrogen separation performance
- Reactor temperature control for peak performance

RESEARCH TO APPLICATION

Research on how to produce hydrogen from natural gas, synthesis gas and carbon monoxide using membrane reactor systems is plentiful [5][8][10]. Most focuses on Steam Methane Reforming (SMR) [5] [8] and Water Gas Shift (WGS) [2] [3] [4] using palladium as the membrane material that allows hydrogen to pass through without other molecules. Research on small modular designs is also plentiful [6] [9]. The problem is, with all this research on how to generate hydrogen and ammonia using new leading edge technology, little has been implemented in an ammonia production facility. The ten metric ton per day (10 MTPD) modular design referred to as the Ammonia Manufacturing Box (AMB) includes many of these new and leading edge designs. The design focus is low capital cost, low energy consumption, safety and operability. The overall objective is to provide low cost fertilizer to the farmer's field produced from locally available raw materials.

AMMONIA 10 MTPD DESIGN

Ammonia is produced in hundreds of plants worldwide and is one of the largest commodity chemical globally at 160 million metric tons. The 1950's was the decade of the 100 – 200 mtpd reciprocal plant. Late in the 1960's and 1970's was the period of the MW Kellogg large centrifugal ammonia plant (600 – 1000 mtpd). Also in the 1970's CF Braun invented the Purifier plant with low energy performance at 1000 mtpd. MW Kellogg was the dominate player in the 1970's and into the 1980's with replicated plants in the 1000 or 1500 mtpd size [2]. Starting the 1980's energy consumption was the primary focus when natural gas prices were increasing all the competitors adjusted their designs to reach a 7.0 Gcal/MT energy point where it stands today. CF Braun had the clear advantage by having a low energy design already in service. With all the ammonia licensors having the same energy consumption level the next attack was on capital cost. The economy of scale on a per ton basis can result in a 30% cost savings if properly applied up to the limit of parallel trains. This lead to ever larger Mega Ammonia plants in the 2000's and 2010's. Today the low end of the capacity range is 2000 mtpd and the largest 3300 mtpd with the cost per daily metric ton of nameplate capacity about $250,000 for the ISBL. The supporting offsites might cost another $125,000 for Brown field while Greenfield $250,000 or even more. Large plants cost more than $US 1 billion but on a per ton basis are considered the most cost effective. Today however the micro plant is economical and cost effective if the latest technology is applied keeping the CAPEX at $500,000 per daily ton for the complete plant and the energy consumption at 7.0 Gcal per ton.

Since 2000 most of the Rice University Chemical Engineering Senior Design Projects (Capstone) have been to design a synthesis gas chemicals plant with various feedstocks, at varying sizes, in remote locations. Sustainably, Green, Process Intensification, Modular, Mobile, Micro Processing Technology, Local designs for Local needs, Renewables, Latest Technology and Creativity were all themes used in making the Capstone project problem statement. Many of those projects sifted through the current research and found the cost effective new technologies around hydrogen production, ammonia production and modular designs.

SMALL MODULAR ADVANTAGE:

The overall objective is reduce the cost of fertilizer to the farmer in remote locations. Mega plants are the low cost producers in most cases but only for those located within a reasonable shipping distance to the end user (farmer). With the focus on the farmer in a remote location the Mega Ammonia plant is just not a good fit. In many locations the transportation of ammonia to site is difficult and costly plus in many poorer countries there are other related costs added to the price before it gets to the farmer's field. For example a Tanzanian farmer might be able to buy ammonia for $500/MT at the plant gate in Qatar but that price then has to be increased to cover ocean and land transportation ($50 est.) to the farmers field and other costs such as import duties, tariffs, currency adjustments, wholesalers and retailers mark ups

($150 est.). The Qatar plant gate price of $500/MT grows to $700/MT or more at the farmer's field. Several countries subside the fertilizer price to the farmer to offset this higher cost and at the same time subsidize food prices. In developing countries food supply and price are important issues along with balance of payments.

The 10 MTPD modular ammonia plant that can be installed for $5 million ($500,000/DMT) operate consuming energy of 7.0 Gcal/MT while using locally available raw materials, natural gas, air and water.

EQUILIBRIUM REACTORS: THE MEMBRANE ADVANTAGE

Ammonia plants have 7 reactor systems: Sulfur Removal, Primary Reformer, Secondary Reformer, High Temperature Shift, Low Temperature Shift, Methanator and Ammonia Converter. Of those 7 all but 2 are good candidates for membrane reactors because they are equilibrium reactions.

SMR is a good choice as it is central to producing hydrogen from reforming natural gas and/or CO_2.

$$CH_4 + H_2O \leftrightarrow CO + 3H_2 \qquad (+\ 206\ kJ/mole)$$

The SMR as a membrane reactor is challenging because the reforming reaction occurs at high temperature (900 C) and a high heat flux is required to feed the highly endothermic reaction (+206 kJ/mole). It has not been ruled out but other options are being considered.

WGS is a good choice as it completes conversion of feedstock to the maximum amount of hydrogen in a range of lower temperatures (450 C to 200 C).

$$CO + H_2O \leftrightarrow CO_2 + H_2 \qquad (-\ 41\ kJ/mole)$$

For the 10 t/d plant this looks to be the best application of a membrane reactor.

The ammonia converter is a good choice as an equilibrium exothermic reactor but at the converter pressure (125 bar) cooling and condensing of the product stream to remove ammonia is preferred as less costly.

$$N_2 + 3H_2 \leftrightarrow 2NH_3 \qquad (-\ 92.44\ kJ/mole)$$

The ammonia converter will be a new micro design that keeps the operation at or close to the equilibrium curve though out its length.

WGS MEMBRANE REACTOR:

Key mechanical features:

a) Removable Bundle
b) Double Tubesheet
c) Partitioned Shellside
d) Membrane Tubes
e) No Internal Flanges
f) Reaction, Separation and Heat Transfer all in one unit replacing several items.

Key process parameters:

a) Pressures in line with existing ammonia plants but with higher pressures where it helps the micro size plant overall efficiency and capital cost.
b) Temperatures closely tracking the equilibrium curve and close approaches in the heat transfer equipment.
c) Flows within the tubes and outside that match the best performance for the reactor conversion, membrane separation and heat transfer.
d) Compositions that complement the rest of the unit. Hydrogen directly to the process nitrogen stream. More complete conversion of CO.

In an ammonia plant there are two shift reactors, high temperature shift (HTS) with a high reaction rate and low temperature shift (LTS) with a close approach to equilibrium at a low temperature consuming most of the CO and producing maximum hydrogen.

Some ammonia plant designers have offered a medium temperature shift (MTS) as a means of lowering the CAPEX of the plant by eliminating equipment items. This has not caught on because the operations within the two shifts HTS and LTS are so different with respect to catalyst performance, type and temperature of operation. The micro plant at 10 mtpd will have a precision temperature shift (PTS) where the temperature is closely controlled along the length of the tube containing catalyst. Plus the catalyst at each temperature step will be suited specifically for that temperature and concentration of reactants.

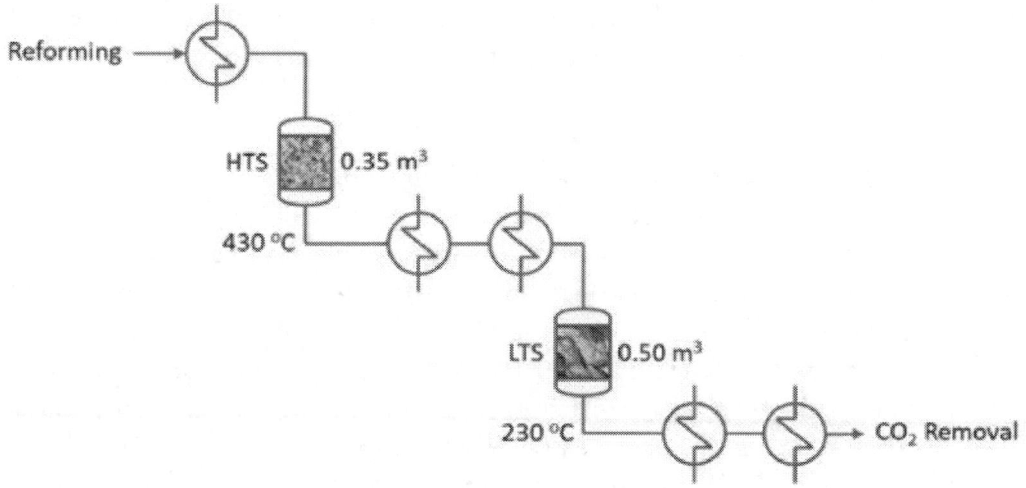

Current PFD at 10 MTPD

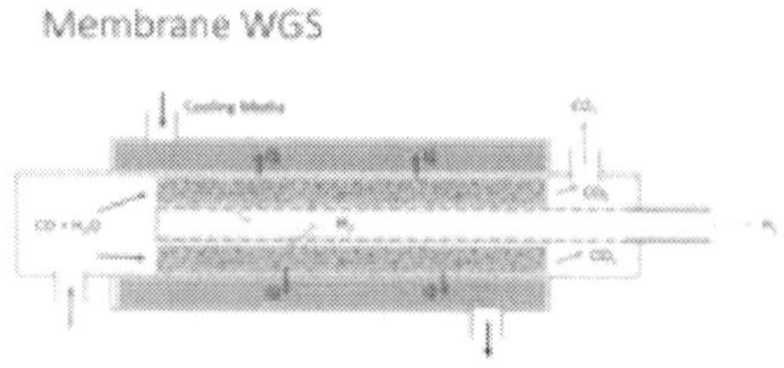

Single Tube Membrane WGS

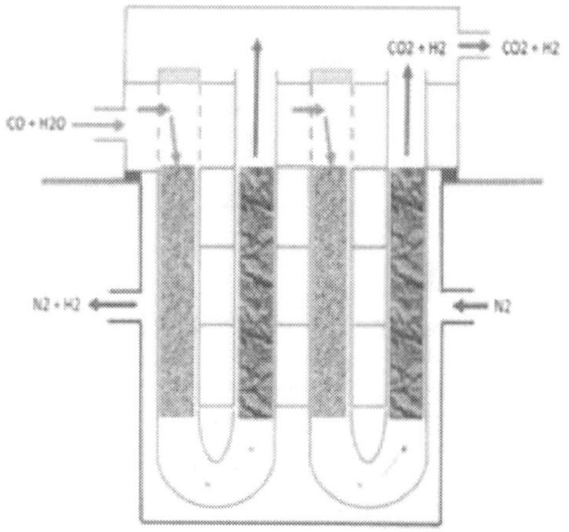

Multi Tube Membrane WGS

Our starting point was two smaller reactors with the same performance to the normal HTS and LTS, but on a micro scale and using modern technology these adiabatic reactors can be combined into one precise temperature controlled membrane reactor containing proportionally less catalyst and a precision temperature match for each catalyst along the reactor tube. Ultimately the new design saves catalyst, equipment and energy for the process. Hot syngas will enter the tube where the reaction will occur rapidly. The heat generated from the reaction passes from the tube through the wall to the nitrogen sweep gas on the shell side. The tube wall is also the membrane allowing hydrogen to preferentially pass through it into the speed nitrogen on the shell side. The shell side of the MWGS contains nitrogen which will be used in the ammonia reaction later. This nitrogen is not contaminated by the hydrogen but actually enriched as it builds up to the required 3:1 ratio required for ammonia production.

MEMBRANE SELECTION:

Palladium is widely studied as the membrane material because it provides the best separation (100% pure H2) but it is very expensive at 1800 – 2800 US$ per ounce. Palladium is the most effective membrane material because it uses proton transfer to allow the hydrogen molecule to disassemble on the tube side and reassemble on shell side with almost perfect separation of the hydrogen at 100% purity. But Pd adds $75,000 - $150,000 to the cost of the small WGS reactor in our small plant. Palladium also performs well at 450 C but at 400 C the permeation rate is cut in half and at 350 C the CO starts to block the hydrogen from permeating through the tube [2]. Nickel has been tried as a lower cost alternative but it is only 8% effective as Pd in separating the hydrogen [3]. Ceramics and zeolites are promising [11]. At this point zeolites are the most promising as being most overall cost effective. The membrane and precise temperature control will allow us to reduce the size of the WGS reactor by 25 – 50%.

MECHANICAL DESIGN:

The WGS takes place in a porous catalyst filled tube with nitrogen flowing on the outside of the tube which sweeps away the permeate hydrogen while cooling the reaction. This u-tube heat exchanger reactor has a removable bundle facilitating simple field change out of the catalyst charge. The u-tubes allow the tubes to grow and shrink without binding and a two tubesheet design allows all the connections to be located at the top of the reactor/separator/exchanger.

PROCESS DESIGN:

The process design takes the research papers into the process plant instead of shrinking an existing design 200 times. Process Intensification is central to the design leading to a modular design that can be shop fabricated and transported to site as a single unit. Also with a one size fits all design it can be replicated a 1,000 or 10,000 times to lower the capital cost per ton of production.

Capital Cost (CAPEX) is reduced at every turn. The catalyst is more effective in our WGS membrane reactor/separator/exchanger. Equipment is combined and now new improved catalyst formulations are possible. But it goes well beyond the WGS into every process portion of the plant including integrated utilizes.

Operational Costs (OPEX) is also as low as 7.0 Gcal/MT or even lower depending upon how far we go with equilibrium reactor designs and other equipment within the plant.

SAFETY AND OPERABILITY:

The operation of the plant will be by farm center operators with extreme instrumentation and monitoring by computers and a minimum of operator attention required. The operator is there to keep things running though maintenance activities and reporting activities and to shut down the plant if required because the instrumentation did not. The operator is not there to make operational moves but rather to shut down and secure the plant if necessary. Safety is built in with a goal of minimum operator interface, maximum monitoring and automatic adjustment and shutdown if necessary to a safe condition.

SUMMARY:

- No significant change to the ammonia process design since the 70s (centrifugal plant) and 80s (low energy plant). Since the 60s and though to today all licensors have been making ever larger plants to take advantage of the economy of scale to build lower cost per ton plants but that has its limits at 2000 – 3000 MTPD.
- The 10 MTPD plant is modular and takes process intensification to its practical limit, using membrane reactors with integral heat and mass transfer.
- Membrane material will be ceramics/zeolites. Palladium would add $75,000 to $150,000 to the cost of the membrane WGS and is considered too expensive.
- The WGS is described as a system ready to be integrated into the plant and beta tested.
- The operation of the plant will be by farm center operators. Experts are brought in as needed.
- Safety will be enhanced because of addition instrument and computer operational control.

REFERENCES

[1] Appl, Max, *Ammonia: principles and industrial practice.* 1999. Print.

[2] Alexander S. Augustine, Yi Hua Ma, Nikolaos K. Kzantzis, "High temperature palladium membrane reactor for high temperature water – gas shift reaction", *Intl. Journal of Hydrogen Energy 36* **2011**, 5350 – 5360 [X002]

[3] K. Babita, S. Sridhar, K.V. Raghavan. "Membrane Reactors for fuel cell quality hydrogen through WGSR – Review of their status, challenges and opportunities." *Intl. Journal of Hydrogen Energy 36* **2011**, 6671-6688

[4] M.D. Dolan, R. Donelson, N.C. Dave, "Performance and economics of a Pd-based planar WGS membrane reactor for coal gasification." *Intl. Journal of Hydrogen Energy 35* **2010**, 10994 - 11003 [X003]

[5] Abu El Hawa, H. W., Lundin, S., et al. "Steam methane reforming in a Pd-Au membrane reactor: Long-term assessment" *Intl Journal of hydrogen energy* **2016** *41*, 10193 – 10201 [C003]

[6] Jensen, Klavs F. "Silicon-Based Microchannel Systems: Characteristics and Applications." *MRS Bulletin* **Feb 2006,** *Vol. 31,* 101-107.

[7] Multhaup, Robert H. and Eschenbrenner, Gunther P., *Technologies Harvest, Feeding a Growing World Population.* 1996. Print.

[8] Simakov, David S. A. and Sheintuch, Mosche. "Experimental Optimization of an Autonomous Scaled-Down Methane Membrane Reformer for Hydrogen Generation." *Ind. Eng. Chem. Res.* **2010,** *49,* 1123-1129.

[9] Vaccaro, S., Malangone, L., Ciambelli, P. "Micro Scale Catalytic Reactor for Syngas Production." *Ind. Eng. Chem. Res.* **2010,** *49,* 10924-10933.

[10] Yan, Y., Cui, Y., et al "Experimental investigation of methane auto-thermal reforming in hydrogen permeable membrane reactor for pure hydrogen production." *Intl Journal of hydrogen energy* **2016** *41*, 13069 – 13076 [C002]

[11] Yonamine, W., Thangavel, S., et.al. "Performance Analysis of a water gas shift membrane reactor for integrated coal gasification combined cycle plant." *Energy Conservation and Management 174* **2018,** 552-564 [C001]

R. B. Strait
Adjunct Professor Rice University = RBS1@rice.edu
Ammonia Consultant = Richard.Strait.Syngas@gamil.com
February 26, 2020

590276 Integrated Mid-Scale Natural Gas Liquefaction Solution- Factors That Can Lead to a Viable Mid-Scale Project

Abstract:

Mid-scale LNG is generally defined as a single or multiple LNG train(s) with capacity of 1.0 to 3.0 MTPA. The main factors supporting this development is a shorter schedule and certainty of cost by using standardized designs. The mid-scale concept is being applied to both the revamp and green field projects.

The Mid-Scale LNG projects are conceptualized to realize benefit of handling manageable pieces thus reducing the risk of cost and schedule overruns. The mid-scale solution not only brings more vendors and contractors available in this space, but it also allows for more design alternates and thus more flexibility during the project execution thereby reducing the overall risk significantly. In some cases, depending on the guiding factors, multiple mid-scale trains can be a potential viable alternate for a LNG project.

The paper discusses the integrated mid-scale LNG solution in detail and its advantages in certain conditions. The integrated solution entails an off the shelf standardized pre-engineered design using the APCI SMR / DMR technology for Liquefaction and Technip Energies design for Pre-Treatment section. The paper discusses few case studies along with risk based economic analysis and illustrates the factors, if realized, that can lead to a viable Mid-scale LNG development in certain cases.

590092 New Modeling Approach of Mixing Quality and Axial Hold-up Distribution in Agitated Liquid-Liquid Flow Mixers for Enabling Process Intensification with Validation Using Advanced Techniques

Abstract:

Liquid-liquid mixing, extraction, reaction, separation, and beneficiation have found a wide range of industrial applications. New theoretical based models have been developed and validated to enable process intensification by properly predicting the mixing quality and the local dispersed phase hold-up distributions in the axial direction of a continuous liquid-liquid flow mixer. The models relate the mixing index and axial hold-up distributions to a Peclet Number, the average dispersed phase hold-up, the input flow rates and the physical properties of the dispersion. The developed models offer a new understanding of the nature of the mixing process in continuous mixers with two liquid-liquid immiscible phases in flow. Experimental work was also performed to investigate the mixing index and axial dispersed phase hold-up distribution using advanced Laser Transmission Technique (LTT) in a mixer setup. The experimental results were used in validating the models. The experimental and the predicted data agreed well for a wide range of impeller speeds/flow ratios. Furthermore, advanced Gamma-ray based visualization measurement techniques can be integrated to validate the model in an intensifying way.

589564 Commercial Experiences and Further Development of WINTRAY® - High-Throughput and High-Efficiency Liquid-Liquid Extraction Column -

Abstract:

Liquid-liquid extraction is a separation process for recovering the target components in the feed by contact with the solvent. Liquid-liquid extraction is a versatile technology with a wide range of commercial applications. It is used in the processing of refinery, petrochemical, coal and biomass-derived chemicals, complex organic compound production such as pharmaceuticals and nonferrous metals production. In addition, liquid-liquid extraction also is becoming important in industrial wastewater treatment due to the increase of environmental regulations in the world.

In designing the process which uses liquid-liquid extraction, the type of extractor needs to be selected from several kinds of equipment. The selection is very important in capital cost because the size of the extractor and required plot area varies by the type of extractor. The type of extractor also effects the required solvent circulation ratio and the energy consumption for the regeneration in the distillation section. In addition, the maintainability varies with type of extractor. These are directly linked to the operation cost of the facility.

The WINTRAY® which is liquid-liquid extraction column, has been developed to obtain a high-throughput and high-extraction efficiency. The throughput is 2-5 times higher than conventional equipment depending on the application. Conventional liquid-liquid extraction equipment is described below to compare with WINTRAY® .

Mixer-settler can achieve high stage efficiencies because of its long residence time. And it can handle high-viscosity liquids. The operation range is relatively wide because the rotation speed of the agitator is adjustable in the field. It is used in petrochemical, nuclear and nonferrous metal fields. But it requires huge plot area especially when a lot of stages are required. In addition, the equipment and maintenance cost are relatively high because it comes with the several number of agitators.

Rotating disc contactor (RDC) is also widely used in petrochemical, chemical, pharmaceutical and nonferrous metal fields. RDC comes with several theoretical stages in a single column with single agitator, therefore the equipment cost, and operating cost are moderate. And the rotating speed of the agitator can be adjustable in the field like as mixer-settler. On the other hand, because its throughput is relatively lower compared with other extraction column, the diameter of the column tends to be huge. And due to the difficulty in preventing the back mixing in the column, scale up and diameter of the extraction column are limited. In addition, the equipment cost is higher than static extraction column because it has agitator.

Sieve Tray which comes with multi perforated-plates, is relatively effective respect to both capacity and extraction efficiency of the extraction column. It also provides the several theoretical stages in single column like as RDC. It is static equipment, therefore, the equipment cost and maintenance cost are relatively lower than those of RDC. The scale up is relatively easy, and it is used in refinery, petrochemical and chemical industrial fields. However, because of the demand for increased petrochemical production and for cost reduction of the extraction process, further improved throughput and extraction efficiency for the extraction column has been desired.

The WINTRAY® which is liquid-liquid extraction column, can obtain a high-throughput and high-extraction efficiency. It combines two kinds of trays, composed of plates and weirs with unique shape openings, arranged alternately. The heavy liquid is dispersed into the light liquid in the column by the openings on weirs. After dispersed phase liquid accumulates on trays, it flows through the openings in each weir and then changes to liquid jets and makes several dispersed drops. The dispersed drops are larger than those formed in conventional columns, and shape is also different. The sequence of drop formation, movement and coalescence is achieved as the dispersed liquid moves from one tray to the next. In addition, higher frequency of the series involving dispersion, movement and coalescence can be attained with closer tray spacing. Therefore, the use of the WINTRAY® extractor results in both higher throughput and higher efficiency at the same time.

WINTRAY® was commercialized to obtain high capacity, high efficiency and fouling resistance in the liquid-liquid extraction. It has been already used in 19 commercial plants worldwide for various kinds of applications and conditions. The interfacial tension and density difference between heavy phase and light phase in commercial experiences cover the wide range of 1-35 mN/m and 20-600kg/m3, respectively. It means that WINTRAY® can be applied to various kinds of liquid-liquid extraction in industrial use. In addition, the problem of fouling has not been reported.

If WINTRAY® is applied for new construction of extraction column, the size of the column becomes compact because of its high throughput and high efficiency. High throughput makes the column diameter smaller, and high efficiency reduces number of trays and height of the column. In addition, as compared to extractors coming with driving device, the simple structure which consist of trays and weirs with openings, leads to low capital cost and maintenance cost.

When revamping the existing extraction columns, throughput can be increased only by replacing the internals with WINTRAY®. And also, the energy consumption in the solvent regeneration in distillation section can be saved by reducing the solvent circulation rate of the extraction column because of the high tray efficiency.

As an example of the performance of WINTRAY, the application to aromatic extraction column is described. In a commercial project, the perforated tray in the existing column was removed and WINTRAY® was installed in the aromatic extraction column uses sulfolane as a solvent. The number of trays and tray spacing were not changed after the replacement. After the modification, the amount of solvent is decreased 10% after installing WINTRAY® , and the BTX loss is reduced by 42% because of higher efficiency than the perforated tray, even though the capacity was increased 17%. In this project, the incensement of the capacity was limited because of the capacity of other facilities around the extraction column, and the actual capacity of WINTRAY® is higher than that. In the pilot test results, the throughput of WINTRAY® is approximately 40% higher than that of sieve tray.

In another commercial project, WINTRAY® was newly constructed in parallel with the existing RDC which is used for aromatic extraction. To reduce the aromatic concentration in the raffinate, the raffinate which contains less aromatic obtained from WINTRAY® , was mixed to the raffinate from the RDC and supplied to the downstream process. When using only RDC, the lower limit of the aromatic concentration in the raffinate is approximately 6%. On the other hand, it was reduced to 3% by parallel operation with WINTRAY® because of its high efficiency. In addition, even though the capacity per column cross-section area of WINTRAY® is double higher than that of the RDC, and the amount of solvent is decreased 20%, BTX loss is saved 70%. The height of the extraction part of the

WINTRAY® is 15% lower than RDC. We herein report the characteristics and review example cases of the usage and the advantages of WINTRAY® in detail.

We also introduce further development to improve the throughput of the extraction column. The newly developed tray is composed of novelty shape plates and weirs with openings.

In addition to the introduction of performance improvement, the prediction method of the flooding velocity is also developed. The flooding velocity is a key factor for deciding the diameter of the extraction column in new construction project, and deciding the capacity in the internal replacement project. The liquid-liquid extraction is a versatile technology with wide range of applications. The feed and operation conditions vary widely, especially in refinery industry. Therefore, the accurate prediction of the flooding velocity is very important when the technology is applied to new application and new conditions. We introduce the comparisons between the experimental results and the results which are estimated by our method.

Realizing the promised benefits of process intensification and integration through proper controller design

Michael A. Taube, S&D CONSULTING, INC., Houston, Texas, USA
Isuru A. Udugama Ph.D., Technical University of Denmark (DTU), Lyngby, Denmark
Brent Young, Ph.D., University of Auckland, Auckland, New Zealand
Christoph Bayer Ph.D., Technische Hochschule Nürnberg (TH-N), Nürnberg, Germany

Introduction

Process integration (mass and energy integration) is predominantly established as means for productivity improvement and cost minimization. The domain expertise is subject to development with active research being carried out into the topic. Currently the concepts of process integration are applied in the design of traditional petrochemical processes [1–3], while research has been carried out to also include process intensification objectives (usually reducing capital and operating costs by combining unit operations, functions or phenomena) to areas such as pharmaceutical manufacturing [4] as well as biotechnology based production processes [5]. Because of this diverse domain of application, process integration and intensification (PII) has been applied in numerous large-scale systems such as industrial (petro)chemical distillation units to less abundant applications such as micro-reactors.

Lutze et al. [6] defined process intensification as "a process development/design option which focuses on improvements of a whole process by adding/enhancing of phenomena through integration of unit operations, integration of functions, and integration of phenomena and/or targeted enhancement of a phenomenon within an operation". There is an increasing interest in application of intensified and multi-functional processes in chemical industry [7]. Several applications of process intensification principles are realized so far on an industrial scale including reactive distillation, micro-reactors and rotating packed bed systems, to name a few.

From a process design point of view, PII enables multiple operations or phenomena to be combined in a single unit operation while providing opportunities for identifying novel more sustainable process alternatives. From an economic point of view, intensified unit operations potentially offer multiple benefits including, but not limited to:

- Reduced capital cost by reducing the equipment required to carry out product synthesis and refinement [8]
- Reduced energy and operations cost due to better utilization of heat and materials flows [9], e.g. in baffled reactors
- Increased production due to the continuous in-situ removal of inert or inhibitory material [5], in case of combining reaction and separation in one-unit operation, e.g. reactive distillation.

Due to the potential benefits of PII, it has been adapted into a variety of commercial applications; examples include side draw distillation columns (SDC), divided-wall column (DWC), reactive distillation columns (RDC) and the in-situ removal of inhibitory molecules in bioreactors. However, despite the potential benefits, design and control of these intensified and integrated processes can be complicated due to loss in degrees of freedom resulting from combining operations, tasks or phenomena. It has been the author's experience that intensified processes in general are poorly controlled, due to the lack of process understanding that is caused by complexities brought on by PII. This paper describes a handful of case studies where additional complexities in terms of process control are outlined. Each case study also explains how the complexities brought on by PII were successfully tackled. The final section deals with lessons learnt that describe in general terms challenges faced and the ways to tackle potential pitfalls of control with respect to PII.

Success Stories

Side Draw Distillation Tower

The first example is a distillation column that performs the final clean-up of a C4 alcohol product at a specialty chemicals plant, removing both a light and heavy impurity from the final product; the purified C4 alcohol is taken as a side stream from the tower. This plant was built in early 1970's. Over time, a few of the original instruments – namely the Bottoms Accumulator Level and Bottoms Flow meter – were either abandoned or removed because of unreliability (due to fouling service). Control structure issues also affected the operations of the tower.

The original controls design on the overheads system had the Overhead Temperature manipulating the very large Reflux Flow, and the Accumulator Level manipulating the very small Distillate Flow (as usually found with "classic refinery control" designs); refer to Figure 1. This configuration never worked as the Reflux Flow was 2.5-3.5 times the Feed Flow, compared to the Distillate Flow which was only 2.5% of the Feed. Thus, the Accumulator Level, Overheads Temperature and Distillate flow constantly oscillated (but not due to poor controller tuning)

Figure 1: Side Draw Tower Original Control Structure

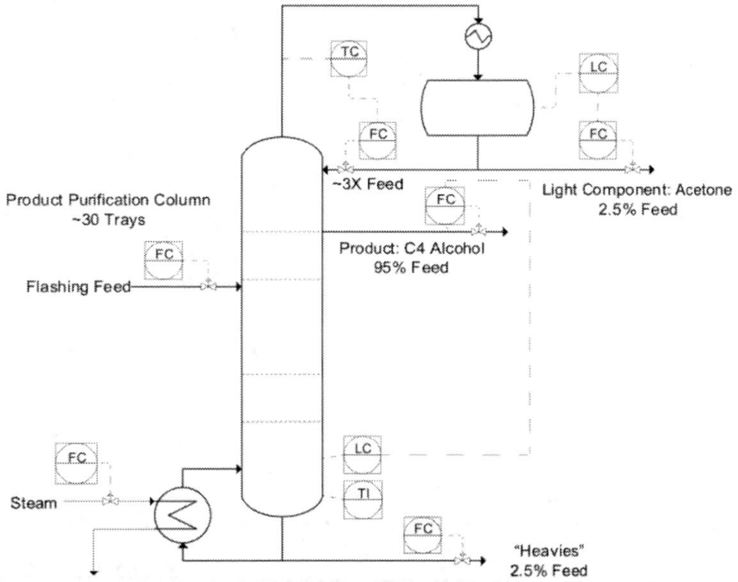

When starting the investigation, the tower was operated in a "semi-batch" style: The Feed was held constant and the overheads and Side Draw controls were operated in fully automatic/cascade control, but with the very poor behavior noted above. The Bottoms, however, was operated manually, as follows: The Bottoms Flow was blocked-in until the Sump Temperature High Alarm was reached (around 300°F); then the Bottoms Flow would be unblocked. The Sump Temperature would precipitously, and with very little delay, fall, at which point the Bottoms Flow would then be blocked back in. This "dump/fill cycle" would be repeated about every 24 hours, as illustrated in Figure 2.

Figure 2: Bottoms Flow Valve & Bottoms Accumulator Temperature

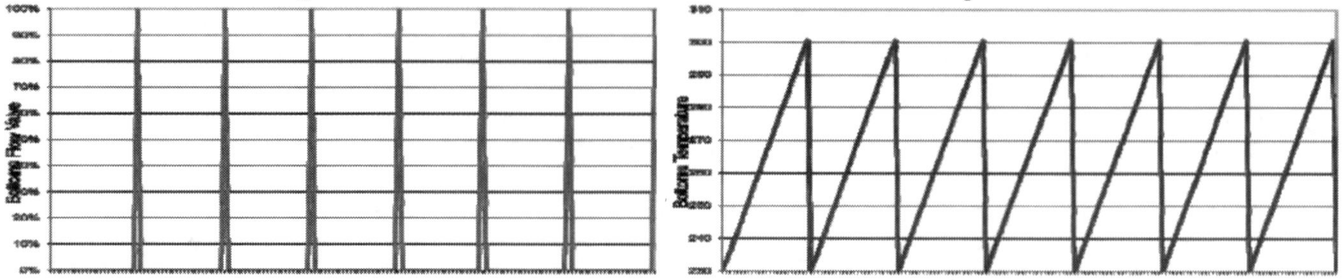

The solution to the stability and control issues was comprised of several elements:

- The first change was to the overheads control, swapping the MV-CV pairing based on an RGA as shown in Figure 3; a conceptual composition profile is also shown. Once tuned, the revised controls stabilized the mass and energy balance on the tower.
- Control Bottoms Flow based on Bottoms Temperature and tune it as an integrating process, based on the observed process response.
- The abandoned instrumentation on the Bottoms Flow stream and Accumulator were also upgraded as part of the controls' effort
- A Reflux/Feed controller was also provided for operator convenience to ensure sufficient heat was provided to the tower for the required fractionation.

- **Figure 3: Side Draw Tower Final Control Structure**

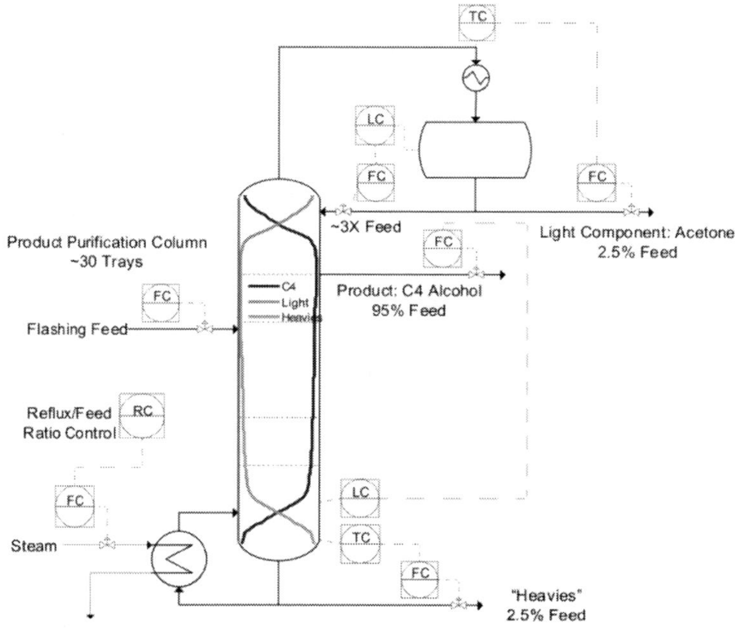

This initial example illustrates that even a simple integrated/intensified process, such as a side draw distillation column, can present challenges to robust and stable control. The analysis performed and the solutions performed indicated the need to understand the process fundamentals when proposing controls to these types of distillation units as "standard" solutions that are used on non-intensified process operations can lead to suboptimal solutions. Finally, this example also shows how a lack of proper instrument selection can create problems with unit operations in general.

Divided Wall Distillation Tower

This example covers a divided wall column (DWC) for solvent recovery. It is well known that DWCs are particularly challenging to design and to control. In this particular case, the level of complexity was raised further through the (apparent) reactivity of the mixture and the deep energetic integration of the column into the overall process.

In addition to the divided wall design, this design is further intensified by the fact that the solvent vapor supplies all the heat to the overall process. Refer to Figure 4. The solvent is Methanol which becomes diluted by water and contaminated by organic trace compounds. The removal of the organic compounds is of paramount importance, since they potentially contaminate the high purity product of the process. However, the objective of the DWC's design was not a sharp split. Remaining water in the Methanol up to a few percent is acceptable. Besides, the organic load of the water to be discharged should be as low as possible due to the cost for disposal and the loss of solvent through the side draw should be minimal due to the cost for solvent replacement.

Figure 4: Divided Wall Column

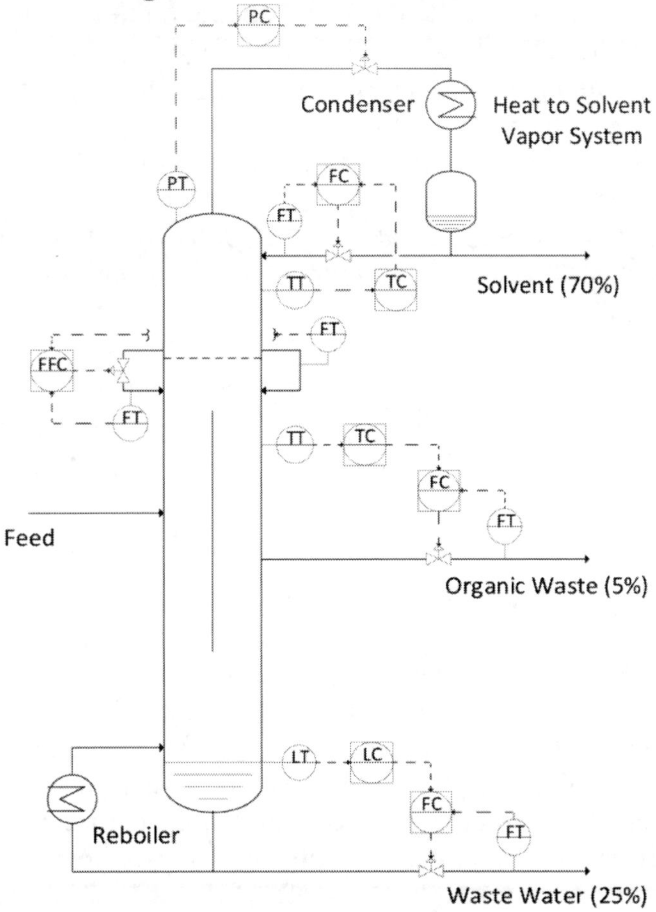

Generally, a DWC is preferred for separation problems with the middle boiler being present in relatively large quantities. Contrary to that, the side draw in this case is in the order of only a few percent (of the Feed flow rate).

The overall control structure is of moderate complexity. The column pressure is fixed by a control valve that manipulates the vapor flow to the condenser. Reflux mass flow is manipulated to control the temperature in the rectification section above the dividing wall. The liquid split ratio between the prefractionation section and the main section (i.e. the section behind the dividing wall) is adjusted by a valve; the operations team manually sets the operating point. The side draw rate is variable so as to fix a temperature reading between the side draw and the upper end of the dividing wall. The bottoms level was controlled by a mass flow controller.

The column was designed such that the overhead vapor would be sufficient to heat most other columns in the process. Hence, the column's reboiler was mostly running at full load. This further ensured that virtually no Methanol ended up in the bottoms and in the side draw.

During operation two significant problems arose:

(1) A light organic compound formed in the column with a boiling point similar to the process' main product. It left the column with the Methanol vapor (the solvent) and subsequently contaminated the product.

(2) At times, it was difficult to keep the column in the desired point of operation. Apparently, intermediate boiling compounds got trapped in the column and accumulated.

Observations and solutions:

(1) The light organic compound was one of the key trace compounds identified during process design. Hence, it was well understood how it formed and how its formation could be reversed. The column was further equipped with an acid dosing system to lower the pH in the prefractionation section. The originating compound, an aldehyde, was lower-boiling and as such easy to remove in the existing process set-up.

(2) Periodically, the accumulating intermediate boiling compounds were released with the Methanol to the solvent cycle. Since these compounds did not deteriorate the quality of the product, no immediate measures were taken. However, the built-up of these intermediate boiling compounds and the sub-optimal operation of the column impaired plant economics. And, as a side effect, the operating teams improved their skill sets by riding through these "periodic storms".

One of the options discussed was the installation of a second side-draw to periodically or continually purge the intermediate boiling compounds from the process and to monitor their built-up rate closely.

This example also illustrates both the great economic benefit and the problems that may arise from highly integrated designs. In case complex designs fail to perform as expected, it is incredibly helpful to have additional nozzles installed that enable the attachment and integration of additional measurement or sampling devices that help to better understand operational issues and facilitate their resolution.

Dairy Processes

Globalization together with the ever-increasing consumer demand for dairy products in recent decades have resulted in the creation of large-scale industrial dairy processing plants [18]. This is especially true for milk powder production where, for example, Fonterra New Zealand installed the world's largest dairy spray dryer which is able to convert 4-4.5 million liters of fresh liquid milk every day [19]. As a result of these large production volumes and high energy requirements, process intensification is a well-established practice in the modern dairy plant.

One area where process control been implemented to address complexities bought on by process intensification in diary production is in the water-removal part of the plant where multiple effect evaporators are common place; some facilities also incorporate Mechanical Vapor Recompression (MVR) to improve the energy efficiency of the operation which is a principle cost driver. There have been a few success stories that have been recorded in connection of controlling intensified evaporation process. In Karimi et al. [20], the authors had success in using inferential cascade control in operating a multi-effect falling film evaporator in a milk powder factory in Iran.

In comparison to the examples that have been discussed above (which are from the petrochemical industry), the dairy industry has some unique process understanding and process monitoring challenges. Firstly, to successfully control an intensified process it is necessary to understand the underlying process from a steady-state, as well as dynamics perspective. In the petrochemical industry, validated process simulations play a major role in this aspect. This is not the case in dairy industry where accurately modelling components in milk is difficult. A promising development in this arena is the creation of Pseudo-milk components for commercial process simulators which, in turn, allows for the modelling of entire dairy processes [21].

Similarly, the lack of monitoring methods available to accurately measure the quality of a production process further complicates the control of intensified (as well as normal) unit operations. Another promising development in this aspect has been the proposal of online monitoring tools such as hyper spectral imaging used in milk powder characterization [22].

In comparison to the petrochemical examples, the research results reported with dairy processes are laying the foundation on which intelligent process control systems can be developed for dairy unit operations. This is because the development of process simulation and monitoring tools lets one understand and identify the underlying phenomena as well as process dynamics which is of critical importance to successfully implement feasible process control solutions.

Lessons Learned

When analyzing these examples, it is clear that better understanding of process engineering fundamentals and process dynamics led to the development/design of better process control structures and/or unit operations designs. Once this understanding was established, the solutions proposed to successfully control the unit operations were relatively simple. This is an important finding as, in many situations, control engineers, as well as control academics, tend to suggest complex process control solutions to intensified distillation units which are both difficult and expensive to implement and also fail to be robust. Analyzing the success stories also allows for the following conclusions to be drawn.

I. In many situations the control in intensified processes is not well thought out during the process design phase
II. Considering the control of intensified processes at the design phase of a unit operation can be beneficial as this will guarantee the unit operation can be controlled at desired production targets once operational
III. Understanding the underlying process dynamics and phenomena prior is a prerequisite to developing stable and robust process control structures for intensified distillation units
IV. Process simulations and plant data visualization play a significant role in accurately identifying the underlying process dynamics and phenomena.

This paper has identified some of the criteria for the successful implementation of IIPs, but , clearly, there is a need for more research and, especially, industrial application of IIPs. Some suggested areas of research and industrial practice include:

- The integrated steady state and dynamic design of large scale plants;
- Further skills development and training of process design engineers;
- Development of new features in engineering and design tools;
- A systematic design and performance assessment & benchmarking; and, most importantly,
- The knowledge feedback and lessons learned from commissioned plants

References

1. Mansouri, S. S.; Sales-cruz, M.; Huusom, J. K.; Gani, R. Systematic Integrated Process Design and Control of Reactive Distillation Processes Involving Multi-elements. Chem. Eng. Res. Des. 2016, Submitted.
2. Stankiewicz, A. I.; Moulijn, J. A. Process Intensification: Transforming Chemical Engineering. Chem. Eng. Prog. 2000, 22–34.
3. Horiuchi, K.; Yanagimoto, K.; Kataoka, K.; Nakaiwa, M.; Iwakabe, K.; Matsuda, K. Energy Saving Characteristics of the Internally Heat Integrated Distillation Column (HIDiC) Pilot Plant for Multicomponent Petroleum Distillation. J. Chem. Eng. JAPAN 2008, 41, 771–778.

4. Roberge, D. M.; Zimmermann, B.; Rainone, F.; Gottsponer, M.; Eyholzer, M.; Kockmann, N. Microreactor Technology and Continuous Processes in the Fine Chemical and Pharmaceutical Industry: Is the Revolution Underway? Org. Process Res. Dev. 2008, 12, 905–910.
5. Heintz, S.; Börner, T.; Ringborg, R. H.; Rehn, G.; Grey, C.; Nordblad, M.; Krühne, U.; Gernaey, K. V.; Adlercreutz, P.; Woodley, J. M. Development of in situ product removal strategies in biocatalysis applying scaled-down unit operations. Biotechnol. Bioeng. 2017, 114, 600–609.
6. Lutze, P.; Gani, R.; Woodley, J. M. Process intensification: A perspective on process synthesis. Chem. Eng. Process. Process Intensif. 2010, 49, 547–558.
7. Nikačević, N. M.; Huesman, A. E. M.; Van den Hof, P. M. J.; Stankiewicz, A. I. Opportunities and challenges for process control in process intensification. Chem. Eng. Process. Process Intensif. 2012, 52, 1–15.
8. Schoenmakers, H. G.; Bessling, B. Reactive and catalytic distillation from an industrial perspective. Chem. Eng. Process. Process Intensif. 2003, 42, 145–155.
9. Ho, T. J.; Huang, C. T.; Lee, L. S.; Chen, C. T. Extended ponchon-savarit method for graphically analyzing and designing internally heat-integrated distillation columns. Ind. Eng. Chem. Res. 2010, 49, 350–358.
10. Harmsen, G. J. Reactive distillation: The front-runner of industrial process intensification. Chem. Eng. Process. Process Intensif. 2007, 46, 774–780.
11. Tuchlenski, A.; Beckmann, A.; Reusch, D.; Düssel, R.; Weidlich, U.; Janowsky, R. Reactive distillation — industrial applications, process design & scale-up. Chem. Eng. Sci. 2001, 56, 387–394.
12. Panda, D.; Kannan, A. Equilibrium and Rate Based Simulation of MTBE Reactive Distillation Column. 2014, 8, 998–1004.
13. Sneesby, M. G.; Tade, M. O.; Smith, T. N. Two-point control of a reactive distillation column for composition and conversion. J. Process Control 1999, 9, 19–31.
14. Pöpken, T.; Steinigeweg, S.; Gmehling, J. Synthesis and Hydrolysis of Methyl Acetate by Reactive Distillation Using Structured Catalytic Packings: Experiments and Simulation. Ind. Eng. Chem. Res. 2001, 40, 1566–1574.
15. Al-Arfaj, M.; Luyben, W. L. Comparison of Alternative Control Structures for an Ideal Two-Product Reactive Distillation Column. Ind. Eng. Chem. Res. 2000, 39, 3298–3307.
16. Georgiadis, M. C.; Schenk, M.; Pistikopoulos, E. N.; Gani, R. The interactions of design, control and operability in reactive distillation systems. Comput. Chem. Eng. 2002, 26, 735–746.
17. Chung, Y.-H.; Peng, T.-H.; Lee, H.-Y.; Chen, C.-L.; Chien, I.-L. Design and Control of Reactive Distillation System for Esterification of Levulinic Acid and n -Butanol. Ind. Eng. Chem. Res. 2015, 54, 3341–3354.
18. Munir, M. T.; Wilson, D. I.; Depree, N.; Boiarkina, I.; Prince-Pike, A.; Young, B. R. Real-time product release and process control challenges in the dairy milk powder industry. Curr. Opin. Food Sci. 2017, 17, 25–29.
19. GEA Building the World's largest dairy Spray Dryer http://www.gea.com/en/stories/building-the-worlds-largest-dairy-spray-dryer.jsp.
20. Karimi, M.; Jahanmiri, A.; Azarmi, M. Inferential cascade control of multi-effect falling-film evaporator. Food Control 2007, 18, 1036–1042.
21. Munir, M. T.; Zhang, Y.; Yu, W.; Wilson, D. I.; Young, B. R. Virtual milk for modelling and simulation of dairy processes. J. Dairy Sci. 2016, 99, 3380–3395.
22. Munir, M. T.; Wilson, D. I.; Yu, W.; Young, B. R. An evaluation of hyperspectral imaging for characterising milk powders. J. Food Eng. 2018, 221, 1–10.

AUTHOR INDEX